Alvin Valentine Lane

Adjustments of the Compass, Transit, and Level

Alvin Valentine Lane

Adjustments of the Compass, Transit, and Level

ISBN/EAN: 9783337185923

Printed in Europe, USA, Canada, Australia, Japan

Cover: Foto ©Lupo / pixelio.de

More available books at **www.hansebooks.com**

ADJUSTMENTS

OF THE

COMPASS, TRANSIT, AND LEVEL.

BY

A. V. LANE, C.E., PH.D.,

ASSOCIATE PROFESSOR OF MATHEMATICS,
UNIVERSITY OF TEXAS.

———∘∘⦂∘⦂∘∘———

BOSTON:

PUBLISHED BY GINN & COMPANY.

1886.

J. S. CUSHING & CO., PRINTERS, BOSTON.

PREFACE.

A N examination of those text-books in which this subject is or should be treated, reveals the fact that it is for the most part very meagrely and arbitrarily presented, sometimes dismissed with the statement that the adjustments should be made by the maker of the instrument. The method, when given, is generally without explanation or proof that it will accomplish the desired object, so that the student must either take the author's word for it, get some one to explain it to him, or work it out for himself; and being usually unprepared for such original work, he is in danger of adopting the first course mentioned, or of leaving the matter in doubt and mystery.

A great source of trouble lies in the fact that such authors are expected to express themselves in accurate terms and do not; the word "half," for example, being so often used for that which is at best but approximately so, that the student marvels at the talisman for such diverse operations being so uniformly that particular fraction.

Perhaps the absence of explanation and proofs of the

correctness of some of the methods is largely due to the
difficulty of making the subject clear to those who have
not studied Descriptive Geometry. It is believed, however, that one whose attainments in the line of mathematics go no further than through Elementary Trigo-
nometry will experience no difficulty with the following
discussion of the adjustments of the three principal instruments used by surveyors and engineers.

This little volume has been called forth by the neea
of such an exposition of the subject, felt by the author
for some time past in presenting the matter to his classes
in Engineering, and any suggestions in the line of
improvement will be acceptable.

A. V. LANE.

University of Texas,
Austin, May, 1886.

ABBREVIATIONS.

———◦◦◦———

A, horizontal axis of telescope (Transit).

B, axis of the level bar (Level).

C, line of collimation (Transit, Level).

H and V, horizontal and vertical plane of reference.

I, intersection of A and C (Transit).

R, plane of revolution of adjusted C (Transit).

S, axis of spindle (Compass, Transit, Level).

W, intersection of cross-wires (Transit, Level).

Y and y, longer and shorter distances from B to the centers of the telescope's wye bearings (Level).

$l.t.$, level tube (Compass, Transit, Level).

$l.t.c.$, level tube-case (Compass, Transit, Level).

$l.t.c.a.$, level tube-case axis (Compass, Transit, Level).

I. THE COMPASS.

——∘∘⦂ℴ⦂∘∘——

1. The Plate-Levels, — *to so adjust them that when their 'bubbles are centered, the plate shall be horizontal.*

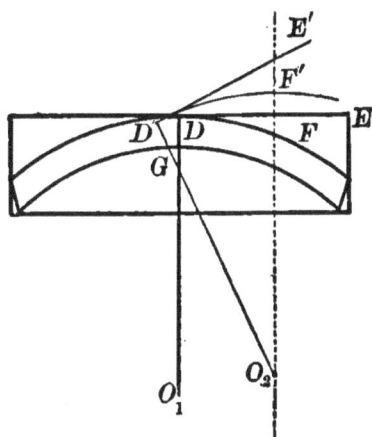

FIG. 1.

Let the *l.t.c.* GE be turned in a vertical plane and about its center G through an angle. The bubble, which was at D, the center of the *l.t.*, will move to a point F' found by raising a vertical through the new position of the center of curvature of the *l.t.* The arc $D'F'$ through which the bubble has moved subtends the angle

$$O_2 = DGD' = \text{the angle between } DE \text{ and } D'E';$$

i.e., the motion of the bubble is proportional to the angle through which the *l.t.c.* is turned about its center.

Let us suppose that the *l.t.c.a.* D_1E_1 is not parallel to the plate JK_1, but makes with it an angle a. Then, when the bubble is brought to the center, D_1E_1 being horizontal, JK_1 will differ a from horizontality. If now the plate

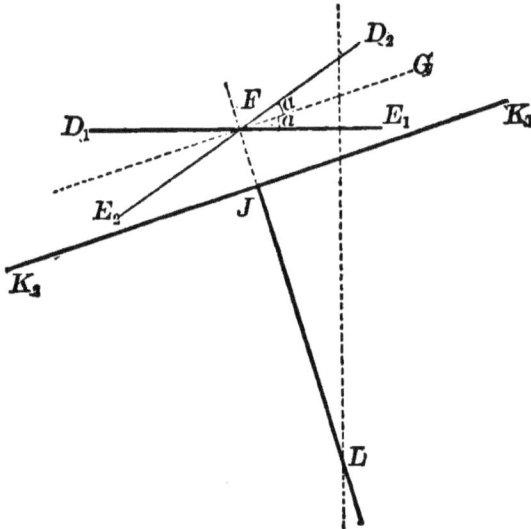

Fig. 2.

be turned 180° on S, the *l.t.c.a.* will have changed ends, and will lie in a vertical plane parallel to the one which before contained it, and distant therefrom by twice the distance of the *l.t.c.a.* from S. Since the two halves of the *l.t.c.* are precisely alike, the effect of this operation on the motion of the bubble is clearly the same as that of a rotation in a vertical plane and about the center of the *l.t.c.* through an angle $2a$. If then the bubble is brought half-way back to the center by the screws at the ends of the *l.t.c.*, and the rest of the way by changing the position of *S*, the *l.t.c.a.* will take the adjusted position *FG* (parallel to the plate), and *S* will be vertical, the plate also horizontal. See Methods of Adjustment, p. 33.

2. The Sights, — *to make them vertical when the plate is horizontal.*

The plate being horizontal, if a vertical line is sighted to and each sight made to range with it from any point of the other, they will be vertical. See Methods.

3. The Needle, — *to so adjust it that any one vertical plane may contain its end points, its center of rotation, and the center of the graduated circle.*

If the ends of the needle do not in every position cut opposite degrees (point to divisions which differ 180°), this adjustment is needed. Failure to do so is due to the point of the pivot not being in a perpendicular to the plane of the graduated circle at its center, or to the needle being bent, or to both causes.

Suppose that, one end of the needle pointing to a division G_1, the other fails to point to its opposite, J_1. Let the pivot be bent until the failure is corrected, when the conditions will be such as are represented in the diagram. Reversing the sights exactly, the points F_1, D_1, E_1, E_1' take the positions F_2, D_2, E_2, E_2'. Evidently the division N_2 is approximately half-way between M_2 and G_2, the angle F_2D_2O differing slightly from F_1D_1O, owing to the displacement of F_1 and F_2 from O. Hence, if the needle be so bent that its end point E_2 cuts half-way back to G_2, its ends will be approximately in line with its center. If the pivot be now bent until the needle cuts the readings J_2 and G_2, F_2 will be at or very near some

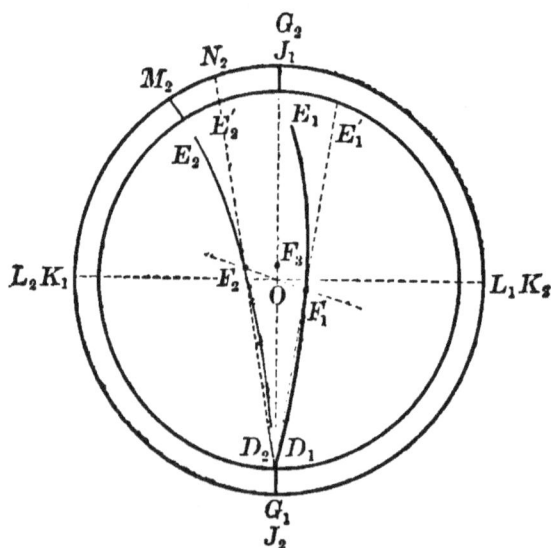

FIG. 3.

point as F_3 in the line G_2J_2. So by repeating this process the ends of the needle will be brought into line with its center, which center will be somewhere in the line G_2J_2, as at F_3.

When this has been done, if one end of the needle be held at K_2 (90° from J_2) and the pivot bent towards

K_2L_2 until the other end cuts L_2, F_3 will be brought into the line K_2L_2 and nearer to its proper position.

By repeating this last operation on the pivot with reference to the lines G_2J_2 and K_2L_2 alternately, its point will be adjusted.

1. The Levels. — Same as I. 1.

2. The Needle. — Same as I. 3.

3. The Line of Collimation, — *to make it perpendicular to the horizontal axis of the telescope, so that, by revolution about that axis, it will generate a plane and not a conical surface.*

It should be remembered that C is the line of sight and is determined by the optical center of the object-glass and the intersection of the cross-wires. The wires should be respectively vertical and horizontal; they may be made so by causing one to range with a plumb-line or known vertical line, since they are perpendicular to each other.

Assume a horizontal circle with center I, and let C_1 differ a from IE, a perpendicular to A. Revolving C_1 about A_1 till it again cuts the circle, it takes the position C_2, differing a from ID, and therefore $180° - 2a$ from C_1. Turning the instrument about S $180° - 2a$, C_2 becomes C_3 in coincidence with C_1, while A_2 takes the position A_3. Revolving C_3 about A_3 until it again cuts the circle, it takes the position C_4, differing a from C_5 perpendicular to A_5. C_5 and ID, respectively perpendicular to A_5 and A_2, differ by the same angle as those positions of the axis; viz., $2a$. Since the arc 4 5 subtends a, $5D$ subtends $2a$, and $D2$ subtends a, the ratio of the arc 4 5 to the arc 4 2 is clearly that of a to $4a$, or one-fourth. If then C_4 were at C_5, one-fourth of the way toward C_2, it would

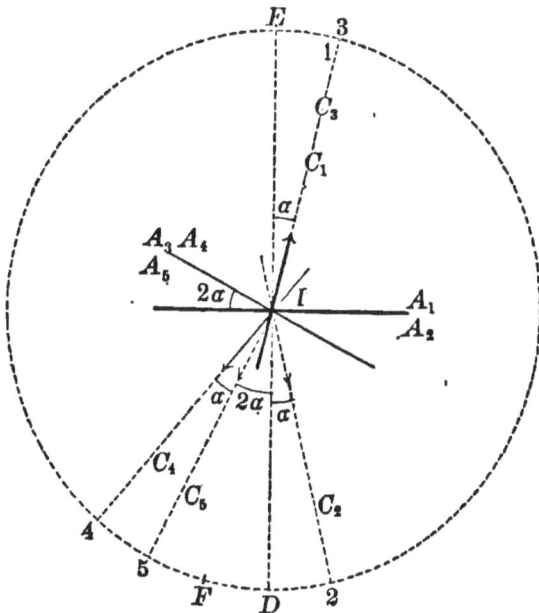

FIG. 4.

be perpendicular to the horizontal axis of the telescope, and hence in adjustment. In practice the above conditions cannot usually be perfectly realized. Thus the points 1, 2, 4, 5 are not in one plane, which introduces a slight error, from the fact that, until adjusted, C revolves about A in a conical surface and so moves to the right or left as it leaves a horizontal position; and again, the points 2, 4, 5 are not taken exactly on a circle, but in a straight line on top of a stake. The errors in each case are small, as the displacements of the points from a horizontal plane and the distances apart of the last three are taken quite small in comparison with their distance from I.

After this has been done, it may be found that the intersection of the cross-wires is not in the center of the field of view of the eye-piece. Since the position of C depends only on the object-glass and W, we may, without affecting it, move the eye-piece by the proper screws until the center of the field of view is brought into coincidence with the intersection of the cross-wires. For the eye-piece is simply a microscope with which we magnify and invert the inverted image formed at the cross-wires where its focus is.

Again, it is evident that, even when C is not perpendicular to A, we may locate three points on the same side of the instrument, which will either be in one vertical plane (in line), if they are all at the same height as I; or very nearly in line, if they do not vary much from that height. Let these points be 1, I and F in the preceding diagram. Setting up the instrument at the middle one I, sighting to 1, and turning C_1 about A_1, it comes to C_2. If C_2 were perpendicular to A_2, the point D would be sighted to. So the point 2 is twice as far from F as the point D upon which to effect adjustment.

, **4. The Standards,** — *to make the bearings of the horizontal axis of the telescope equally distant from the plate,
so that when the instrument is levelled, the line of collimation will, by revolution about the horizontal axis of the
telescope, generate a vertical plane.*

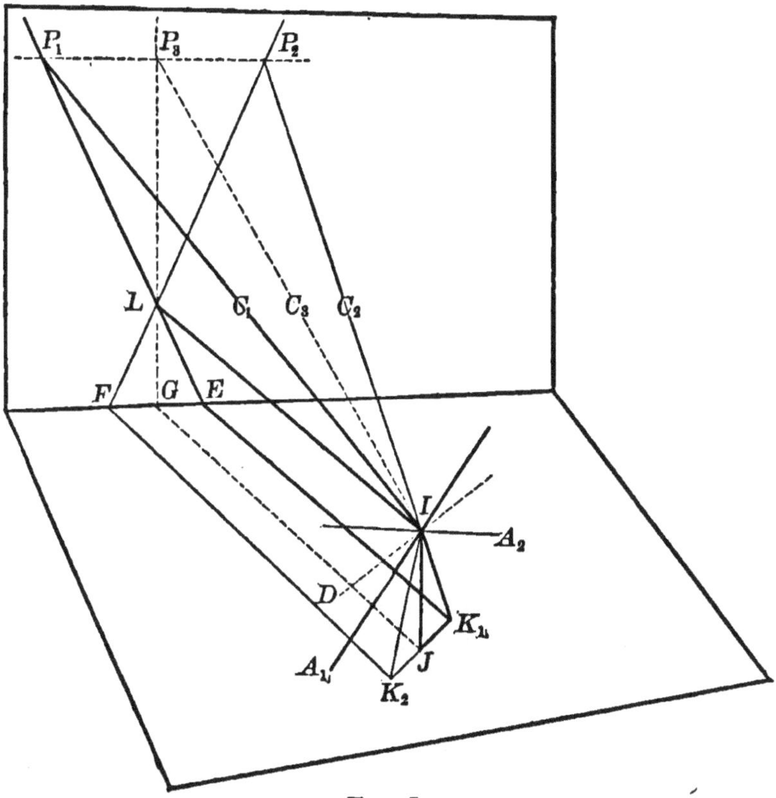

Fig. 5.

Suppose the instrument set up on H, level, and with the
preceding adjustments made.   Suppose A horizontal and
occupying a position ID parallel to H, but not necessarily
so to V.   Draw the vertical IJ and, through J, the line

K_1K_2 in H parallel to ID. Through J draw JG perpendicular to K_1K_2; and, through G, GP_3 perpendicular to the intersection of H and V. Now R, being perpendicular to A or ID, is perpendicular to H, and must cut V in a line perpendicular to the intersection of H and V. Hence R must contain IJ, JG, and GP_3. Take L in GP_3 so that $GL = JG$. C may be directed to L, when it will be parallel to H and perpendicular to the plane DIJ. Since the arm DI of the right angle DIL may be rotated about IL into the position A_1, A may take the position A_1 in the plane DIJ, and C be still fixed upon L. Since, if turned into that plane, C will take the position IK_1, perpendicular to A_1, R, containing K_1 and IL, must now cut H in a line K_1E parallel to IL and therefore to JG, and so must cut V in EL.

If now the instrument be turned on S or IJ through $180°$, A_1 takes a position A_2, making the same angle with IJ as before. A_2 being in the plane DIJ, C may still be directed to L or, in that plane, to K_2. JK_2 being equal to JK_1, and the triangle JIK_1 having turned into the position JIK_2. R must therefore now cut H in a line K_2F, parallel to JG, and V in a line FL.

Draw P_1P_2 in V and parallel to the intersection of H and V. With the axis in the position A_1, C could be directed to P_1; with the axis in the position A_2, to P_2; and in the adjusted position DI, to P_3. The wires should therefore be moved toward P_1 by an amount P_2P_3.

Since JG bisects K_1K_2, it bisects EF, and therefore

GL bisects P_2P_1. So the wires should be moved half-way from their second position, P_2, toward their first, P_1, by changing the height of one of the standards.

Practically, the ground and a vertical wall take the place of H and V. (Given last, under Methods, II. 4.)

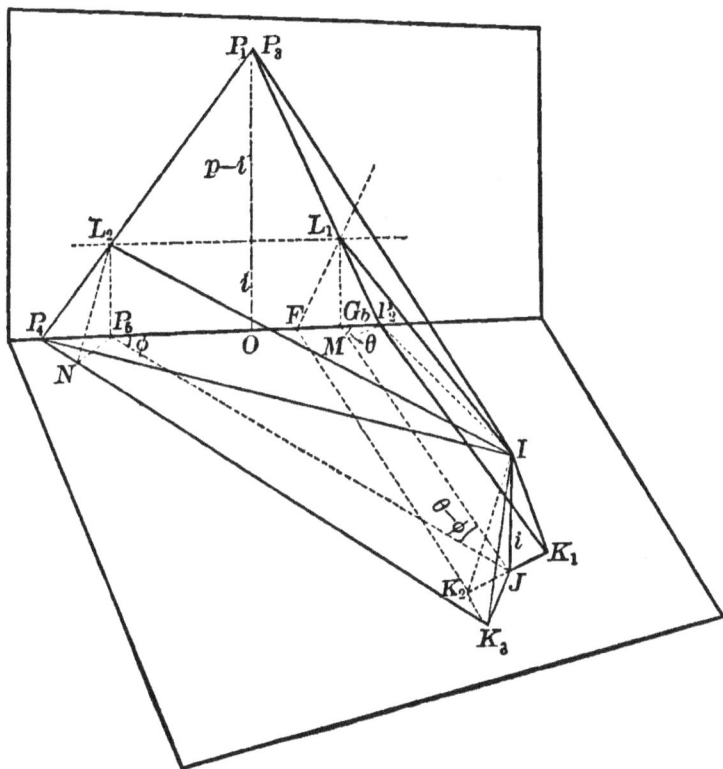

FIG. 6.

Suppose the instrument at IJ, the right-hand end of A the higher, and R occupying the position $P_1P_2K_1$. Imagine the instrument turned 180° on S, so that the triangle IJK_1 occupies the position IJK_2, and R the position

L_1FK_2, just as in the preceding case. Directing C to L_1, imagine it turned to the left, remaining horizontal, until a point L_2 is reached, such that, on revolving C about A, it will again cut P_1. In order for L_2 to be such a point, it is evident that the triangle IJK_2 must have revolved into a position IJK_3, such that perpendiculars JP_5 and K_3P_4 to JK_3 shall meet the intersection of H and V in points P_5 and P_4, respectively on a vertical through L_2 and on the prolongation of P_3L_2.

C was first turned down from P_1 to P_2, then from P_3 to P_4, A lying in the plane IJK_3. If A became horizontal in that plane, R would occupy the position L_2P_5J. So the wires should be moved from P_4 to P_5, by changing the height of one of the standards. We desire therefore to determine the ratio of P_4P_5 to P_4P_2.

Refer the point J to P_2 by $JG = d$, angle $JGP_2 = \theta$, $GP_2 = b$. Let $L_2P_5 = L_1G = IJ = i$. Let $P_1O = p$, and the angle $K_3P_4P_5 = $ angle $JP_5G = \phi$. Then the angle $GJP_5 = \theta - \phi$; also

$$P_5N = JK_3 = JK_2 = JK_1 = MP_2 = b\sin\theta.$$

By similar triangles

$$\frac{P_4P_5}{P_4O} = \frac{i}{p} = \frac{P_2G}{P_2O}. \tag{1}$$

By alternation

$$\frac{P_4P_5}{P_2G} = \frac{P_4O}{P_2O},$$

and this, by composition, gives

$$\frac{P_4P_5 + P_2G}{P_2G} = \frac{P_4O + P_2O}{P_2O},$$

or

$$\frac{P_4P_5 + b}{P_2G} = P_4P_2 \cdot \frac{1}{P_2O} = P_4P_2 \cdot \frac{i}{p \cdot P_2G}, \quad \text{by (1)}.$$

Whence

$$P_4P_5 = P_4P_2 \cdot \frac{i}{p} - b. \tag{2}$$

In the triangle NP_4P_5 we have

$$\tan \phi = \frac{b \sin \theta}{\sqrt{\overline{P_4P_5}^2 - b^2 \sin^2 \theta}}; \tag{3}$$

and in the triangle P_5GJ

$$\frac{P_5G}{d} = \frac{\sin (\theta - \phi)}{\sin \phi} = \frac{\sin \theta}{\tan \phi} - \cos \theta;$$

whence, by (3),

$$\frac{P_5G}{d} + \cos \theta = \frac{\sqrt{\overline{P_4P_5}^2 - b^2 \sin^2 \theta}}{b}. \tag{4}$$

$$P_5G = P_4P_2 - (P_4P_5 + b) = P_4P_2 \cdot \frac{p - i}{p},$$

and

$$P_4P_5 = \frac{P_4P_2 \cdot i - bp}{p}, \quad \text{by (2)}.$$

Substituting these in (4) and clearing of fractions, we have

$$b \cdot P_4P_2 (p - i) + bdp \cos \theta$$
$$= d\sqrt{\overline{P_4P_2}^2 \cdot i^2 - 2bip \cdot P_4P_2 + b^2p^2 - b^2p^2 \sin^2 \theta}.$$

Squaring, cancelling, and taking out the common factor P_4P_2, we find

$$P_4P_2 = \frac{2\,bdp\,[di + b\,(p-i)\,\cos\theta]}{d^2i^2 - b^2\,(p-i)^2}.$$

Substituting this in (2), we obtain

$$\frac{P_4P_5}{P_4P_2} = \frac{i}{p} - \frac{d^2i^2 - b^2\,(p-i)^2}{2\,dp\,[di + b\,(p-i)\,\cos\theta]},$$

the required ratio.

If θ be $90°$ and $p - i = d$, this ratio becomes

$$\frac{i}{p} - \frac{i^2 - b^2}{2\,ip},$$

or, neglecting the small fraction $\dfrac{b^2}{2\,ip}$, it becomes $\dfrac{i}{2p}$.

These conditions are usually very nearly satisfied, so we may consider $\dfrac{i}{2p}$ quite an approximate value of the true ratio. Assuming $i = 5$ ft. and $p = 50$ ft. as about average values, this ratio equals $\dfrac{1}{20}$; while, even if b could equal 6 inches, the neglected fraction $\dfrac{b^2}{2\,ip}$ would only equal $\dfrac{1}{2000}$. (Given second, under Methods, II. 4.)

Suppose that, having as before located the point P_2 from the point P_1, we turn the instrument about half-way around on S and, revolving the telescope, fix the wires again on the lower point P_2. Now, raising them to P_4 at height of P_1, we wish to determine the ratio of P_4P_5 to P_4P_1.

For, as before, R having first the position $P_1P_2K_1$, on turning exactly half-way around, it would take the position L_1FK_2; and, directing C to L_1, we must turn the

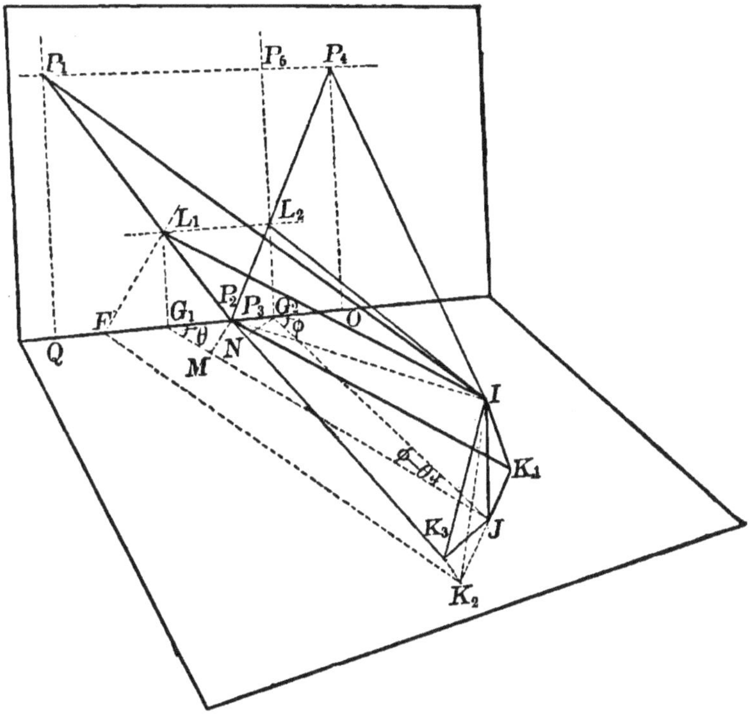

FIG. 7.

instrument to the right until C strikes a point L_2, such that, on revolving about A, the point P_2 is again cut.

The triangle IJK has now revolved into the position IJK_3, and R into the position $P_4P_3K_3$; while, if A became horizontal, R would have the position P_5G_2J.

The references of J to P_2 are now $JG_1 = d$, angle $JG_1P_2 = \theta$, $G_1P_2 = b$.

The angle JG_2O is ϕ, and the angle $G_1JG_2 = \phi - \theta$.

Since, by similar triangles,

$$\frac{OG_2}{P_3G_2} = \frac{p-i}{i}, \quad \frac{P_4P_5}{P_4P_1} = \frac{OG_2}{OQ} = \frac{p-i}{i} \cdot \frac{P_3G_2}{OQ};$$

and we will first determine $\dfrac{P_3G_2}{OQ}$.

$$\frac{P_3G_2}{OP_3} = \frac{i}{p} = \frac{P_3G_1}{P_3Q}. \tag{1}$$

By alternation,

$$\frac{P_3G_2}{P_3G_1} = \frac{P_3O}{P_3Q},$$

and this, by composition, gives

$$\frac{P_3G_2 + P_3G_1}{P_3G_1} = \frac{P_3O + P_3Q}{P_3Q},$$

or,

$$\frac{P_3G_2 + b}{P_3G_1} = OQ \cdot \frac{1}{P_3Q} = OQ \cdot \frac{i}{p \cdot P_3G_1}, \quad \text{by (1)}$$

Whence

$$P_3G_2 = OQ \cdot \frac{i}{p} - b. \tag{2}$$

In the triangle NG_2P_3 we have

$$\tan \phi = \frac{b \sin \theta}{\sqrt{P_3G_2^2 - b^2 \sin^2\theta}}, \tag{3}$$

and, in the triangle G_1G_2J,

$$\frac{G_2G_1}{d} = \frac{\sin(\phi - \theta)}{\sin\phi} = \cos\theta - \frac{\sin\theta}{\tan\phi};$$

whence, by (3),

$$\cos\theta - \frac{G_2G_1}{d} = \frac{\sqrt{\overline{P_3G_2}^2 - b^2\sin^2\theta}}{b}. \qquad (4)$$

$$G_2G_1 = P_3G_2 + b = OQ \cdot \frac{i}{p},$$

and

$$P_3G_2 = \frac{OQ \cdot i - bp}{p}, \quad \text{by (2)}.$$

Substituting these in (4) and clearing of fractions, we have

$$bdp\cos\theta - bi \cdot OQ$$
$$= d\sqrt{\overline{OQ}^2 \cdot i^2 - 2bip \cdot OQ + b^2p^2 - b^2p^2\sin^2\theta}.$$

Squaring, cancelling, and taking out the common factor OQ, we find

$$OQ = \frac{2bdp}{i} \cdot \frac{d - b\cos\theta}{d^2 - b^2}.$$

Substituting this in (2), we obtain

$$\frac{P_3G_2}{OQ} = \frac{i}{p} - \frac{i}{2dp} \cdot \frac{d^2 - b^2}{d - b\cos\theta} = \frac{i}{p} \cdot \frac{d^2 - 2bd\cos\theta + b^2}{2d(d - b\cos\theta)}.$$

Therefore

$$\frac{P_4P_5}{P_4P_1} = \frac{p - i}{i} \cdot \frac{P_3G_2}{OQ} = \frac{p - i}{2dp} \cdot \frac{d^2 - 2bd\cos\theta + b^2}{d - b\cos\theta},$$

the required ratio.

If θ be 90°, neglecting the small fraction

$$\frac{(p-i)\,b^2}{2\,d^2 p},$$

it becomes

$$\frac{p-i}{2p} = \frac{1}{2} - \frac{i}{p},$$

or approximately $\frac{1}{2}$, since $\frac{i}{2p}$ is usually about $\frac{1}{20}$; while the neglected fraction, under the preceding conditions, is $\frac{1}{18000}$. (Given first, under Methods, II. 4.)

This last is the best and most generally used method. The first is excellent, if divisions on the horizontal circle such as the zeros may be depended on as exactly opposite. In the second, the amount through which the wires are to be moved for correction is such a small part of a small distance that it is difficult of exact application.

5. *a.* **The Vertical Circle's Vernier Zero,** — *to so adjust it that when the vertical circle's zero is in line with it, the line of collimation shall be horizontal.*

If, instead of a full circle firmly attached to A, there is only an arc of a circle, which may be turned upon and clamped to it, as is the case in some instruments, this adjustment depends upon and is made after that of the level attached to the telescope. The method of making it in that case is explained further on (after 6).

Let us suppose that the vernier zero Z_1, is not in its proper position (vertically below the center of the vertical circle) ; so that, when the circle zero z_1 is made to coincide with it, C has the position C_1, making the same angle with the horizontal O_1D_1 that O_1z_1 makes with the vertical O_1F_1.

Let P_1 be some point in C_1. Turning the instrument 180° on S, C_1, P_1, z_1, Z_1 move to C_2, P_2, z_2, Z_2. The

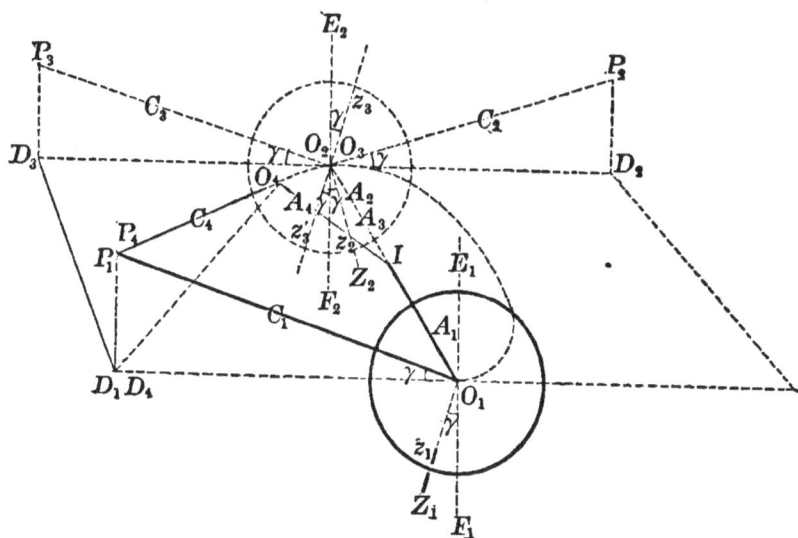

FIG. 8.

angle $P_2O_2D_2$ being equal to γ, if C_2 be turned on A_2 through $180° - 2\gamma$, it takes the position C_3, and z_2 the position z_3, the angles $P_3O_3D_3$ and $E_2O_3z_3$ being each equal to γ. The division z_3' opposite z_3 is now γ from F_2, and 2γ from Z_2. Hence if Z_2 were moved half-way to z_3' and then z_3' brought into coincidence with it, C_3 would take the horizontal position O_3D_3, and, the zeros being also in coincidence, the adjustment would be effected. If A_3 were turned into the position A_4, and C_3 directed to P_4, the angle $P_4O_4D_4$ being slightly different from γ, the ratio corresponding to the above would not be exactly one-half, yet by taking the distance of P_1 from the instrument sufficiently great in comparison with A, the point P_1 may be used for P_3 with but little error; and it is so used, the process being repeated until the whole error becomes inappreciable.

6. *a.* **The Level attached to the Telescope,** — *to so adjust it that when its bubble is at the center, the line of collimation shall be horizontal.*

A full vertical circle being present, and its zero in coincidence with the vernier zero after the preceding adjustment, if the bubble is brought to the center by means of the nut at either end of the level, the adjustment is effected.

5. *b.* **The Level attached to the Telescope.** — If, however, the instrument has the movable arc, the level is first adjusted as follows :

Let D, J_1, E, and J_3 be four points in line, and in the order named; also equally distant horizontally. Suppose the instrument set up at J_1, and C_1 directed to a graduated rod held on the stake D, giving a reading F_1D. Turning the instrument 180° about S, without otherwise disturbing

C_1, C_2 will cut the rod held on the stake E at a reading F_2E, the points F_1, F_2 being at the same height, since C_1 and C_2 make equal angles with the vertical I_1J_1, and

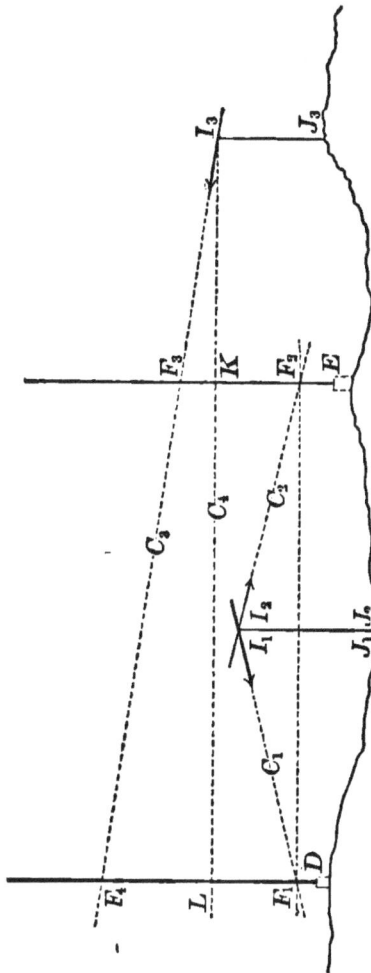

FIG. 9.

the positions of the rod are equidistant from I_1. Let the instrument be now set up at I_3, and, with C in a position C_3, the readings F_3E and F_4D noted. Draw I_3KL parallel to F_2F_1, and therefore horizontal.

$F_4F_1 = F_4D - F_1D$ and $F_3F_2 = F_3E - F_2E$ being known, let $x = LF_1 = KF_2$, and we have, by similar triangles,

$$\frac{F_4F_1 - x}{F_3F_2 - x} = \frac{LI_3}{KI_3} = \frac{3}{1};$$

whence

$$x = \frac{3\,F_3F_2 - F_4F_1}{2}$$

is known.

So, by directing C to a reading $DF_1 + x$, it takes the horizontal position C_4, and the level on the telescope may be adjusted by bringing its bubble to the center, by means of the nut at either end.

6. b. The Vertical Arc's Vernier Zero. — The vertical arc may now be turned on A and clamped with its zero in coincidence with the index zero, effecting its adjustment.

For convenience of application as to sign, etc., it is well to take the first reading just above the top of the higher of the stakes D, E, and to set up at J_3 so that I_3 shall be higher than F_1, F_2; taking the readings F_3E, F_4D greater than F_2E, F_1D respectively.

By driving the stakes at D and E so that their tops are at F_1 and F_2, the first two readings reduce to zero, and F_3F_2, F_4F_1 are the only readings taken; they are used as stated above, x being itself the reading of the target at F_1 for horizontality.

III. THE LEVEL.

———∘∘⦂⦂∘∘———

1. The Line of Collimation, — *to so adjust it that it will pass through the centers of the circles of the wye bearings of the telescope.*

This adjustment is a first step toward making C perpendicular to S.

Fɪɢ. 10.

Let D and E be the centers of the circles of the wye bearings and C_1 the position of C sighting to some point as P_1, distant P_1P_3 from the line ED produced. If now C_1 be turned 180° about ED, it will take the position C_2, making with ED an angle equal to that made by C_1, and sighting to some point as P_2 at the same distance as P_1 from O and P_3. Clearly now for C_2 to coincide with ED, W_2 must be brought to W_3, half-way back to W_1, its former position, by moving it until C_2 takes the position C_3 sighting to a point P_3, half-way from P_2 towards P_1. This may be best accomplished by adjusting each wire in succession, if they be much in error, by means of some line sighted to, until each is nearly right, then completing the adjustment of each. This amounts to substituting for P_1 a line through P_1 and perpendicular to the plane of the paper, for P_2 a like but imaginary line, and for W_1, W_2, and W_3 one of the cross-wires, also perpendicular to the plane of the paper and through those points.

As already explained in II. 3, the center of the field of view should now be brought into coincidence with W by moving the proper screws, and the correctness of the centering may be tested by turning the telescope in the wyes, when the object should not appear to shift its position.

2. The Level, — *to make the level tube-case axis parallel to the line of collimation.*

In the explanation of I. 1 it was shown that in the operation of reversing the *l.t.c.a.* with reference to any plane or line to which its ends were referred, the bubble moves from the center in the first position to some point in the second position, such that the arc between the center and that point is double the arc through which the bubble should be moved back in order to make the *l.t.c.a.* parallel to the line or plane of reference. And this, too, whether that line or plane be horizontal or not.

Since now C contains the centers of the circles of the wye bearings, by reversing the telescope end for end as to the wyes, we may effect this adjustment; the ends of the *l.t.c.a.* being referred to C. But there is a disturbing element to be considered in this connection.

The *l.t.c.* has a screw at one end for lateral adjustment, to bring the *l.t.c.a.* into one plane with *C*, and another, at the other end, for vertical adjustment to make these lines equidistant throughout. If we remove the clips from the wyes, and reverse the telescope end for end, it may be that we do not put it down with the vertical screw immediately below *C*, as before, but slightly rotated to one side. We must therefore examine the effect on the bubble of a slight rotation of the *l.t.c.a.* about *C*, with a view to ascertaining how much if any of the bubble's motion on reversal is due to this cause.

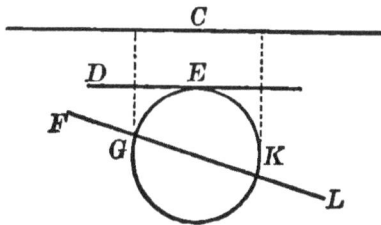

Fig. 11.

Imagine the plane of the paper vertical, and containing *C* and the *l.t.c.a.* Draw *DE* tangent to the *l.t.* at *E*. The highest point of the *l.t.* being that of contact with a horizontal tangent plane, it is evident that, *C* being horizontal, this highest point will lie on a cross-section of the *l.t.* at *E* for any revolution less than 90° about *C*, whether the *l.t.c.a.* is equidistant from *C* or not. For its position does not affect the symmetry of the parts of the *l.t.* on each side of *E*, but simply their amounts.

Suppose C is not horizontal. After a revolution of 90°, G_1G_2' and K_1K_2' turning about G_2' and K_2' respectively, into a horizontal position, G_1 will be at the same height as G_2', and therefore higher than K_1 will be (same height as K_2'). So the bubble will, as the rotation begins, start towards the end G_1.

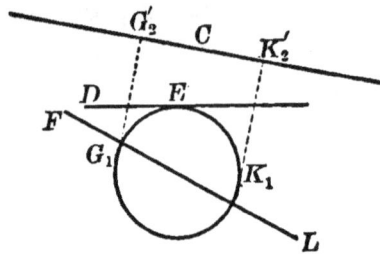

FIG. 12.

Again, in Fig. 11, suppose L to be slightly in front of the plane of the paper, and F a like amount behind. If we rotate towards us, L rises and F falls, the bubble therefore running toward K, instead of remaining at E, as before; while, if we rotate the other way, the reverse occurs. Also, in Fig. 12, if we make a like supposition, since the bubble before moved toward G_1, whichever way rotation took place, while the effect of this new supposition tends to make it move towards K_1 or G_1 (according as we rotate towards us or away), it is clear that these causes may combine to move the bubble one way, or may oppose and even neutralize each other.

Thus it is evident that, knowing nothing of the actual state of the positions of C and the *l.t.c.a.*, we cannot predicate anything as to how much of the motion of the bubble on reversal may be due to the cause just examined. Nevertheless the lateral screw may be, *by trial*, so adjusted as to eliminate the effect of this slight rotation, and finally to bring the *l.t.c.a.* into the same plane with *C*, as in the method given farther on.

3. The Wyes, — *to so adjust them, as to the distances of the centers of their circular bearings from the axis of the level-bar, that the axis of the spindle shall be perpendicular to the line of collimation.*

B is supposed to be made perpendicular to *S*, so that if *Y* be made equal to *y*, *C* will be parallel to *B*, and therefore perpendicular to *S*. But if *B* differs by some angle β from perpendicularity to *S*, by reason of bad construction or some strain received, *Y* and *y* must be given such values as to counteract this and make *C* perpendicular to *S*.

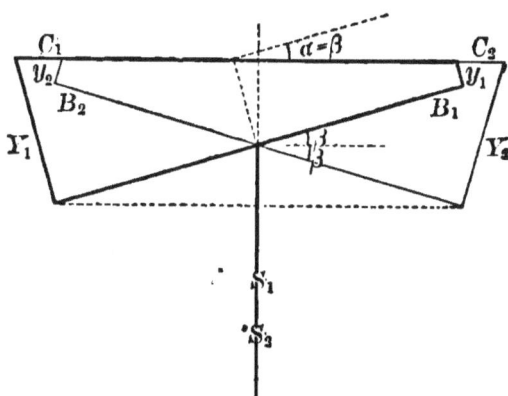

Fig. 13.

From the diagram it is easily seen that the necessary condition is $a = \beta$, with opposing effects on the bubble (a special case of which is $a = 0$, $\beta = 0$, when C is parallel to B as well as perpendicular to S), or, in other words, C making the same angle with B that B does with a perpendicular to S, and S making an acute angle with that half of B which carries Y.

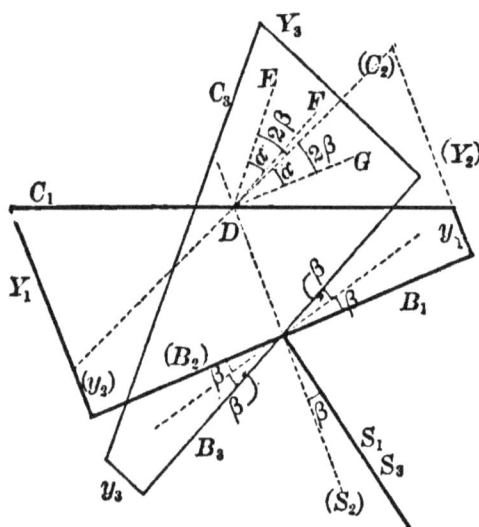

Fig. 14.

Let us suppose S to have a position S_1, making an acute angle $90° - \beta$ with the y half of B. We see that, by reversing about S_1, B_1 and C_1 take the positions B_3 and C_3; while if S_1 had the position (S_2) perpendicular to B_1, B_1 and C_1 would take the positions (B_2) and (C_2), on reversal. Rotating (B_2) through 2β into the position B_3, (C_2) rotates through an equal angle into the position

C_3. Since, then, C_3 makes an angle 2β with (C_2), and (C_2) makes 2α with C_1, C_3 makes $2\alpha + 2\beta$ with C_1.

This may also be shown thus: Drawing DE parallel to C_3 and DF parallel to B_3, we have $EDF = \alpha$. But $FDG = 2\beta$, since FD is parallel to B_3 and DG to B_1. Therefore the (acute) angle between DF and (C_2) equals $2\beta - \alpha$. Hence the (acute) angle between DE and (C_2) or C_3 and (C_2) equals $\alpha + 2\beta - \alpha = 2\beta$; whence the (acute) angle between C_3 and C_1 equals $2\alpha + 2\beta$.

Thus we see that, on reversing about S_1, the bubble moves from the center toward Y over an arc corresponding to $2\alpha + 2\beta$, the effects of α and β being in this case cumulative. Remembering that only 2α of this is due to C's not being parallel to B, and that, if the bubble were moved back over α, C_3 would be parallel to B_3, we see that if we shorten Y until the bubble comes half-way back to the center, we move it over $\frac{1}{2}(2\alpha + 2\beta) = \alpha + \beta$; i.e., β more than the proper amount (α) to make C_3 parallel to B_3. Thus we have replaced the error α by another, β, and made that wye which was the longer, now the shorter; having decreased Y too much for parallelism of C_3 and B_3. Since Y and y have interchanged ends, S now makes an acute angle with the Y half of B, and the angle between C_3 and B_3 is equal to that between B and a perpendicular to S, — precisely the conditions of Fig. 13 (except that S is not vertical). So that the adjustment is effected with B not perpendicular to S.

Suppose the effects of α and β on the bubble are opposing, that α is greater then β (S making an acute angle with the Y half of B). Reversal about S_1 brings B_1 and C_1 into the positions B_3 and C_3; while reversal about (S_2) perpendicular to B_1 would have brought them into the positions (B_2) and (C_2). Introducing the effect of β,

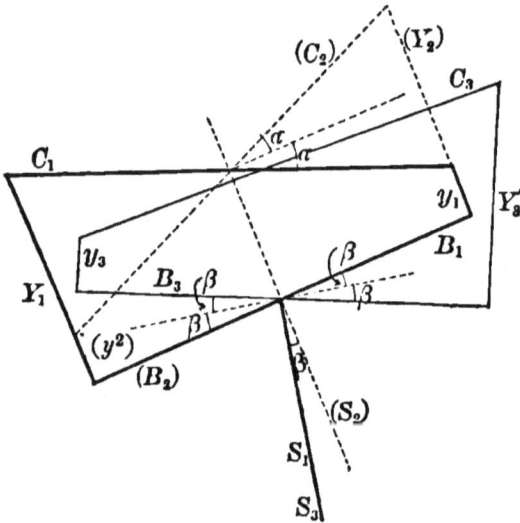

Fig. 15.

(C_2), which had separated from C_1, returns toward but not to it, since $\alpha > \beta$. The combined effect of α and β is thus to move the bubble over an arc corresponding to $2\alpha - 2\beta$ toward Y_3. Since, in bringing it half-way back to the center, it is moved over $\alpha - \beta$ (β less than the proper amount to make C_3 parallel to B_3), there remains an error β between C and B. But since Y_3 has not been made shorter than y_3, S is still making an acute angle with the Y half of B, and so the adjustment is effected, as in Fig. 13.

Finally, suppose α and β opposing, and $\beta > \alpha$. The effect of α being to make the bubble run toward the (Y_2) end of (C_2), and β causing (C_2) to turn back through C_1 to C_3 (making the bubble run away from the Y_3 end of C_3), their combined effect will move it $2\beta - 2\alpha$ toward the Y_3 end of C_3. To move it half-way back we must now lengthen Y_3 (or shorten y_3), and so increase α by $\frac{1}{2}(2\beta - 2\alpha)$; *i.e.*, by $\beta - \alpha$, making the angle between

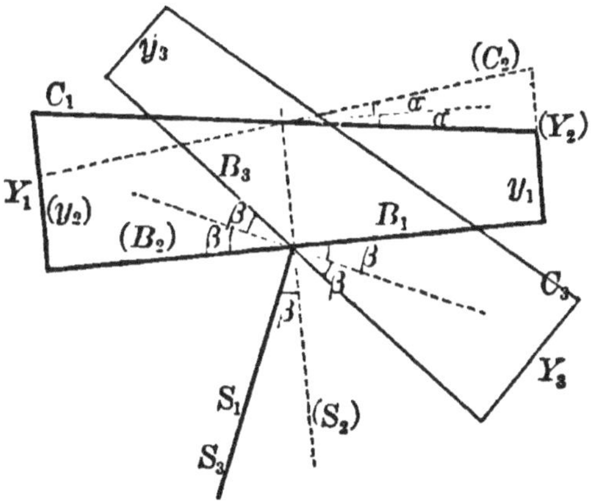

Fig. 16.

B_3 and C_3 $\alpha + (\beta - \alpha) = \beta$, S still making an acute angle with the Y half of B, — again the conditions of Fig. 13.

Thus we see that, when α and β are cumulative in effect, the process results in making them opposing, and if they are at first opposing, they result opposing. For, in the first case, we make the longer wye the shorter, thus introducing a change; while in the latter, although the length of one of the wyes is changed, the longer one remains the longer. In each case a is changed to the amount β (there being no facilities for changing β), and so they are left equal and opposing.

It should be noted that, while C is made perpendicular to S, by thus bringing the bubble half-way back to the center, it must be brought the rest of the way back by the leveling-screws; then C will be horizontal, and if it is made so in two intersecting positions, S will be vertical.

METHODS OF ADJUSTMENT.

———∘o⦂ꙮ⦂o∘———

I. THE COMPASS.

1. The Levels. — Set up the instrument, and bring the bubbles to the center by pressure of the hands on the plate. Reverse the sights, and if the bubbles remain at the center, the levels are in adjustment. If they do not, bring each half-way back to the center by means of the screws at the ends of the level tube-case, the rest of the way by means of the plate, and repeat. Or perform this operation with one level at a time until it is nearly adjusted, then with the other, finally completing the adjustment of each and seeing that both will reverse correctly and remain in the center during an entire revolution of the plate.

2. The Sights. — Observe through the slits a good plumb-line and if either sight fails to range with it, make it do so by whatever means the instrument calls for, — usually, filing a little off of one side of the surface of contact of the sight with the plate.

3. The Needle. — If the needle will not in various positions cut opposite degrees, this adjustment is needed.

Having removed the glass top of the compass-box, with a splinter of wood bring one end of the needle to any prominent graduation, as the zero, having the eye nearly in the plane of the graduated circle, and see if the other end corresponds to the opposite division. If not, bend the center-pin (pivot) with a small wrench, about one-eight of an inch below its point, until that other end of the needle will cut the other zero. Reverse the zeros but not the ends of the needle, and, holding with the splinter the same end of the needle at the new zero, note what division the other end cuts. Bend the needle until that end cuts a division half-way back to the adjacent zero. This puts the needle's ends very approximately in line with its center. Bend the center-pin again, until the ends of the needle will cut the zeros. Repeat until perfect reversion is obtained.

Bring one end of the needle to the 90° division, and if the other end does not cut the opposite division, bend only the pivot until it does, and repeat, using alternately the line of zeros and that of the 90° divisions, until it will cut opposite degrees in any position.

II. The Transit.

1. The Levels. — Adjust by reversal as in I. 1, accomplishing the reversal by means of the readings of the horizontal circle and moving the plate by means of the leveling-screws.

2. The Needle. — Same as I. 3.

3. The Line of Collimation. — Having the instrument set up and leveled on tolerably level ground, make the wires respectively horizontal and vertical by loosening the proper screws and turning the ring around until the vertical wire may be made to coincide with some known vertical line as a plumb-line, or the vertical edge of a building from two to five hundred feet distant. Make the screws tight again.

Select or locate a point from two to five hundred feet distant, clamp the plates, revolve the telescope on its axis, and locate a point on a stake on the other side of and at the same distance from the instrument as the first. Unclamp the plates and turn about the spindle until the wires can be again fixed on the first point. Clamp the plates and again revolve the telescope on its axis. If the intersection of the wires strikes the second point, the line of collimation is in. adjustment. If not, the intersection of the cross-wires should be brought one-fourth of the way back to this second point by means of the pair of cross-wire screws, on the sides of the telescope, which move the vertical wire.

The operator, in loosening one of these screws and tightening the other, should remember that they have their bearings in the cross-wire ring, that a non-inverting telescope inverts the relations of the cross-wires to the object, and *vice versa*. He must therefore, in a non-inverting telescope, proceed as if apparently to move the vertical wire in the opposite direction from that desired.

Test by repetition.

If, when this has been done, the intersection of the cross-wires is not in the center of the field of view, move the latter until they are, by means of the screws which control the eye-piece, loosening and tightening them in pairs; the movement being now direct or as it appears it should be.

Another method is to locate three points in line and all on the same side of the instrument. Then setting up over the middle point, sight to one of the end ones, and clamping the plates, revolve the telescope to sight to the other end one. If the intersection of the wires fails to strike it, move that intersection half-way to the point by means of the vertical wire, as just explained, and repeat. Then center the eye-piece as in the preceding method.

4. The Standards. — Select a tolerably level piece of ground in front of a tall spire, tower, or like object, that shall afford from top to base a long range in a vertical direction. Set up and level the instrument, so that it will be about as far in front of the structure as its telescope is below a good sight-point near the top. Clamp to the spindle, and, fixing the wires on the point selected, clamp the plates and lower the wires to some point found or marked at the base of the structure. (If the ground is not very level, take the point in the face of the building, and at about the height of the ground on which the instrument stands.) Unclamp the plates, and, turning the instrument about half-way around, revolve the telescope, and again fix the wires on the lower point. Clamp the plates and raise the wires to the height of the upper point. If they cut it, the standards are in adjustment. If they do not, bring them half-way to it, by raising the right-hand end (or lowering the left) of the horizontal axis of the telescope, if the wires are to the right of the point; by raising the left (or lowering the right), if they are to the left. Most instruments have a means of making this adjustment at one end of the horizontal axis, — a movable bearing. If the instrument has no such means, file equally a little off of the feet of the higher standard. Repeat until the adjustment is perfected.

Another method is to sight to an upper point, and lowering the wires, fix a lower point at the base of the structure just as before; but, on turning the instrument about half-way around and revolving the telescope, fix the wires again on the *upper* point. Clamping, and lowering the wires to the height of the lower point, if they do not strike it, move them back towards it over that fractional part of the distance, which the height of the instrument above the ground is of double the height of the upper point above the ground. This is to be done by raising the right-hand end (or lowering the left) of the horizontal axis, if the wires strike to the left of the point; raising the left (or lowering the right), if they come to the right. Repeat until the adjustment is perfected.

Another method (which is dependent on the accuracy of the graduation of the horizontal circle) is as follows: Setting up and leveling the instrument, clamp the plates at zero or some other convenient reading, and fix the wires on an elevated (or depressed) point of sight, clamping to the spindle. Unclamp the plates, and, turning through exactly 180°, as shown by the horizontal circle reading, clamp again. If on now raising (or lowering) the wires to the point, they cut it, the standards are in adjustment. If they do not, bring them half-way to it, by changing the height of one of the standards in such a way as to raise the right-hand end (or lower the left) of the horizontal axis, if the wires strike to the right of the point; raising the left (or lowering the right), if they come to the left.

5. a. The Vertical Circle. — (If only an arc is present, see 5 *b*.) Bring its zero into coincidence with the zero of the vernier attached to the standards, and with the telescope find or place some point or horizontal line cut by the horizontal wire and about two or three hundred feet distant. Turn the instrument about half-way around, revolve the telescope, and fixing the wires upon the point, or the horizontal wire upon the line first selected, clamp the telescope and note if the zeros are again in coincidence. If not, loosen the screws that attach the vernier to the standards, and, moving it so as to bring its zero half-way to the vertical circle's zero, make it secure again. Repeat until no error can be detected. Instead of moving the vernier zero, the circle zero may be brought into coincidence with it, and the horizontal wire moved back over half the amount by which it has been thus displaced, provided the error is so slight as not to appreciably throw the intersection of the cross-wires out of the center of the field of view as previously adjusted. Repeat as before.

6. *a*. Level on Telescope. — Level the instrument carefully, and with the clamp-and-tangent movement to the horizontal axis, bring the zero of the vertical circle into coincidence with the vernier zero, and, by means of the screws at each end of the level, bring its bubble to the center, taking care not to jar the instrument out of level.

5. *b*. Level on Telescope. — (When the instrument has only a vertical arc and not a full circle.) On tolerably level ground stake four points in line and equidistant (about 100 feet), calling them (say) D, J_1, E, J_3, consecutively. (It is well that D should not be lower than E, nor J_3 than J_1.) Set up the instrument at J_1, and direct the line of sight to a graduated rod held, as nearly as may be, vertical on the higher of stakes D, E (say D), taking a small reading d_1. Clamp the telescope, and turning the instrument around, take the reading e_1 on E. Set up and level the instrument at J_3, so that the height of the telescope shall be greater than that corresponding to the readings d_1, e_1. Unclamp the telescope and direct the line of sight so that, without changing its position, readings e_2, d_2, respectively greater than e_1, d_1, are obtained on the rod held successively on E and D. From three times $e_2 - e_1$ take $d_2 - d_1$ and divide the result by 2. Set the target at a reading greater by this calculated amount than d_1, and, holding the rod on D, bisect the target by the line of sight and clamp the telescope. By means of the screws at the end of the level, bring its bubble to the center, and see if the line of sight still bisects the target. If it has been jarred out of position, put it back and again bring the bubble to the center. When the line of sight bisects the target and the bubble is in the center, the adjustment is complete.

A simplification of the proceeding is as follows: Instead of taking the first two readings, drive stakes so that their tops are cut by the line of sight, the telescope being clamped. Their tops are then in the same horizontal line. Making the line of sight as nearly horizontal as possible by estimation, take a reading on the nearer of these two stakes, and then, holding the rod on the farther one (target at same reading), if the line of sight

does not bisect the target, turn the telescope by the
tangent-screw so that it will; repeating this until it will
bisect the target held on the far stake at the same reading
as on the near one. Then bring the bubble to the center
by the screws at the ends of the level-tube as before.

6. *b*. The Vertical Arc. — With the bubble of the level
on the telescope at the center, clamp the vertical arc to the
axis of the telescope, with its zero in coincidence with
that of the vernier, and it is in condition to correctly
measure vertical angles.

III. The Level.

1. The Line of Collimation. — Set the tripod firmly,
remove the wye-pins from the clips, so that the telescope
may be turned in its bearings, and, by means of the
leveling and tangent screws, bring either of the wires
into coincidence with a clearly marked edge of some
object, from two to five hundred feet distant. Then
turn the telescope half-way around in the wyes, so that
the same wire may be compared with the edge selected.
If it now coincides with the edge, it is in adjustment;
if not, bring it half-way to it by moving the capstan-
head screws at right angles to the wire in question,
remembering the inverting property of the eye-piece.
Repeat until it will reverse correctly. Then adjust the
other wire in the same manner; or, if their errors are

great, make them nearly correct before exactly adjusting either.

When this has been effected, unscrew the covering of the eye-piece centering-screws, and move each pair in succession so as to bring the center of the field of view to coincide with the intersection of the cross-wires, testing the centering by revolving the telescope in the wyes, when the object should not appear to move. The screws in this case are to be moved as it *appears* they should be. Replace the covering of the screws.

2. The Level-Bubble. — Having the plate about horizontal, place the telescope over either pair of leveling-screws, and, clamping the instrument, remove the wye-pins and bring the bubble to the center by means of this pair of leveling-screws. Reverse the telescope end for end as to the wyes, replacing it with the level-tube immediately beneath, and note whether the bubble remains at the center. Now rotate the telescope slightly to each side, and see if this causes the bubble to move toward either end. If, after the reversal it was at the center, and remained so during the slight rotation, the level-bubble is in adjustment.

If the rotation causes it to move from its position after reversal toward either end, adjust, *by trial*, the horizontal screw so that this will not be the case. Repeat this reversal, etc., until no further adjustment of the horizontal screw is needed. Then (the telescope being in the second position) bring the bubble half-way to the center by means of the vertical screw-nuts at one end of the level tube-case, and repeat the whole process until the bubble will remain at the center after reversal and slight rotation.

3. The Wyes. — Having the telescope in its normal position as to the wyes, place it over a pair of the leveling-screws and bring the bubble to the center by means of them. Turn the instrument half-way around on the spindle, and, if the bubble runs toward either end, bring it half-way back by the wye-nuts on either end of the bar and the rest of the way by the pair of leveling-screws. Then place the telescope over the other pair, and proceed in the same way, changing to each pair successively until the adjustment is completely effected for each, so that the bubble will remain at the center during an entire revolution of the telescope on the spindle.

www.ingramcontent.com/pod-product-compliance
Lightning Source LLC
Chambersburg PA
CBHW031811090426
42739CB00008B/1243